Using the Force and Support Costing System

An Introductory Guide and Tutorial

James H. Bigelow

Manuel J. Carrillo

H. G. Massey

Adele R. Palmer

Prepared for the
Office of the Secretary of Defense

National Defense Research Institute

RAND

This manual steps through a hypothetical analysis as a way to introduce potential users to the interface and procedures used by the Force and Support Costing (FSC) System. Its models and databases are designed to help analysts project the cost implications of proposed changes in defense forces, infrastructure, and assets. The system is being developed by the Forces and Resources Policy Center, a component of RAND's National Defense Research Institute, a federally funded research and development center sponsored by the Office of the Secretary of Defense (OSD), the Joint Staff, the unified commands, and the defense agencies. The system is designed for the Director of Program Analysis and Evaluation in OSD.

This document is addressed primarily to experienced cost analysts. We assume readers are familiar with commonly used cost terminology, databases, and methods, and want to learn how to take advantage of the FSC System in their studies. Terms and acronyms are defined here only as needed to clarify the text and illustrations.

The FSC System user interface is a customized application of Microsoft Excel (currently using Excel Version 5.0 for both Windows and the Macintosh). This document assumes the reader is familiar with the use of menu bars, buttons, scroll bars, and other features of the Excel Graphical User Interface. An understanding of Excel programming codes is not required.

Readers who have installed the FSC System on their computer workstations may use this manual as part of a "hands-on" tutorial in basic FSC operations. Such users can also take advantage of the online FSC documentation (currently in development). Readers who do not have the system may nonetheless find this manual informative as an introduction to some of the FSC System's capabilities.

The illustrations in this manual are from the Fall 1997 version of the system, which included databases and models for Army and Air Force forces and equipment, and for personnel in all Services. To allow this manual to be published as an unclassified document, the illustrations use an artificial database created for demonstration purposes; it covers the period 1995–1999. However, readers should find that they can replicate the procedures illustrated here using more-recent data.

Documentation for the FSC System consists of reports detailing database preparation and online "help" files explaining the methods and interface. Much of the documentation is currently in progress.

Two documents on data preparation procedures have been published previously:

- James Bigelow and Adele Palmer, "Force Structure Costing System: Items in the Army Equipment Database," PM-425-OSD, May 1995

- James Bigelow, Adele Palmer, and Mary Layne, "Force Structure Costing System: Initial Processing of the TAEDP Data Extract," PM-446-OSD, August 1995.

CONTENTS

FIGURES

TABLES

The Force and Support Costing (FSC) System is a set of models and databases that help analysts project the cost implications of proposed changes in defense forces, infrastructure, and assets. This manual uses a hypothetical analysis to illustrate how to use the system.

The illustrated study projects effects on defense costs due to eliminating an active Army division from the force structure. The FSC System allows the user to view the force structure in the current Army program, select a division to be cut, and specify when the cut will occur. The system then translates that deactivation into reductions in personnel and equipment assets, and costs out the implications of using fewer people and less equipment. The system generates summaries of the cost effects, as well as detailed output reports for each affected category of personnel and equipment.

The illustrated study is merely an example. In addition to stepping through the specific procedures for that study, this manual notes other ways the FSC System can be used to analyze the cost effects of various force, asset, and infrastructure policy actions. For example, users can activate or deactivate individual force units (e.g., battalions and companies) in various Services[1] at various times (from now to 30 years in the future).

Readers who have installed Microsoft Excel and the FSC System on their computer workstations may use this manual as part of a "hands-on" tutorial in basic FSC operations. The illustrations in this manual are from the Fall 1997 version of the FSC System (which used Excel 5.0) and may differ somewhat from later versions. Moreover, to allow this manual to be published as an unclassified document, the illustrations use an older, artificial database created for demonstration purposes. However, the general interface design for the FSC System remains constant over time, and readers should find that they can replicate the procedures illustrated here using more-recent system versions and data.

---

[1]Army and Air Force analyses are supported in the current FSC version, but Navy and Marine Corps are planned for future development.

ACKNOWLEDGMENTS

The authors gratefully acknowledge the long-term contributions of current and past members of the FSC project team: Deena Benjamin, Thomas Blaschke, Mary Layne, and Daniel McCaffrey. Deena Benjamin and Thomas Blaschke particularly contributed to an early version of this document.

We also acknowledge the support and encouragement of previous and current RAND managers, Bernard Rostker and Susan Hosek, and the early encouragement of John Morgan, former director of the Office of Program Analysis and Evaluation, who played a central role in shaping the design goals for the FSC System and reviewing its development. A sequence of PA&E project monitors also offered key insights and support: Lawrence Angello, Daniel Barker, and Lance Barnett.

In addition, the document benefited significantly from Mark Y. D. Wang's comments as reviewer.

ACKNOWLEDGMENTS

AK/HI   Alaska and/or Hawaii.

ALO   Authorized level of organization. Indicates extent to which requirements in a unit will be filled during peacetime; ALO 1 is fully manned and equipped.

basops (BOS)   Base operations

compo   Component. Common Army term referring to a variable that identifies whether a unit is in the Active, National Guard, or Reserve components. Sometimes applied to components in other Services in FSC terminology.

CONUS   Continental United States

CSS   Combat Service Support. An Army term for logistics functions such as supply, maintenance, transportation, personnel support, etc. Also refers to units assigned to echelons above division (EAD) that perform these functions.

EAD   Echelons above division. An Army term for force structure assigned to Corps- or Army-level organizations.

EUR   Europe

FSC   Force and Support Costing

FYDP   Future Years Defense Program

NDC   Non-Divisional Combat. An Army term for combat units (e.g., air defense, artillery, combat engineers) assigned to echelons above division (EAD).

OCONUS   Outside Continental United States

ORF   Operational readiness float. Proportional increase in equipment authorizations to provide for repair activities within units.

OSD   Office of the Secretary of Defense

POM	Program Objective Memorandum. The proposed future years' program that each Service submits annually to OSD around May.
POM period	The years covered by the POM. For example, the POM submitted in May 1996 covered fiscal years 1998 through 2003.
RCF	Repair cycle float. Proportional increase in equipment authorizations to provide for repair activities at depots.
SRC	Unit (Standard) Requirements Code
SWA	Southwest Asia
TDA	Table of Distributions and Allowances
TPSN	Troop Program Sequence Number. A five-digit Army code assigned to each unit that identifies to which division or separate brigade the unit belongs or, if not in a division or separate brigade, whether it is an NDC or CSS unit.

- **Action**: A record of a particular set of changes ("deltas") in forces, assets, or infrastructure that are specified as part of a cost exercise ("case"). A case typically includes several actions created by entering, for example, deltas in the number of Army units, corresponding deltas in Army enlisted and officer end-strengths, and corresponding deltas in Army equipment of various kinds.

- **Baseline**: The yearly quantities of forces or assets reported in the FSC program database. Also refers to the annual cost estimates for the assets and operations in the current program.

- **Case**: A particular combination of assumptions, databases, and methods that together generate a single, internally consistent set of cost results. A number of alternative cases may be generated and compared as part of a study concerned with a given topic or issue.

- **Deltas**: Year-by-year differences (positive or negative) between amounts of forces, assets, or infrastructure in the current Budget and POM and alternative amounts specified for the purposes of a cost study. This term is often used by cost analysts to avoid the ambiguity in the word "changes." Whereas "changes" may refer to the movement in a quantity from one year to the next, "deltas" always refer to differences between alternative conditions in a given year.

- **Excursion (or target)**: The yearly quantities of forces or assets that would result from applying deltas to the baseline quantities. Also refers to the annual cost estimates for the assets and operations in the excursion.

- **Method**: A set of modeling techniques and assumptions applied to estimate cost effects for a given action. For example, a simple way to analyze a cut in enlisted personnel is to apply an average cost factor to the change in end-strength; an alternative method is to model the way average costs change over time as the grade and length-of-service profile of the enlisted force adjusts to a cut in accessions. The FSC System lets you choose between these methods, as well as impose additional conditions or constraints.

- **Study**: A general cost exercise to analyze several alternative ways of achieving some specified goal. A study normally includes multiple cases.

Chapter One

INTRODUCTION

The Force and Support Costing (FSC) System enables analysts to project the cost implications of proposed changes in defense forces, infrastructure, and assets. This manual introduces you, the analyst, to the FSC System, using a hypothetical case to show how you might proceed through a cost exercise.

PURPOSES OF THIS MANUAL

This manual serves a combination of purposes. One is simply to acquaint you, as a member of the defense cost analysis community, with the overall design and growing content of the FSC System. Because the FSC System is still expanding, this manual cannot fully document all of the latest features and databases.[1] Instead, it describes the system's range of capabilities, and illustrates them with specific examples.

The second purpose is to encourage you, as a potential or new user, to take advantage of the FSC System's extraordinary flexibility. With this tool, you can analyze cases ranging from small to large changes in forces, equipping, manning, and support.[2] You can postulate specific changes in programmed forces and assets, or work with notional force structures, such as a "typical" infantry division. You can use cost factors[3] provided by the system database or edit the factors to match values from other sources, and you can observe how the estimated cost effects differ. You can even tailor the analysis sequence to fit your own needs as an analysis progresses, perhaps by doing one "what if?" example and then going back to modify data or methods to observe the implications. In addition to illustrating a specific analysis, this manual lays the foundation for exploratory analysis by describing the FSC System's capabilities to support the generic tasks and modeling choices involved in many cost studies.

---

[1]Users who have access to the FSC System can get current information for each FSC version via an online "help" menu.

[2]At the time of this writing, the system can analyze changes in Army or Air Force forces or equipment in any component, and personnel changes in any Service and component. Future plans call for adding modules for Navy and Marine Corps forces and equipment, and for detailed infrastructure modeling.

[3]"Cost factors" (or simply "factors") is a common cost-analysis term used to refer to cost rates, prices, and physical ratios or relationships. Examples are annual pay rates for personnel, and fuel usage per flying hour for aircraft.

Finally, our third purpose is to provide a "hands-on" tutorial for new users who have operational access to the FSC System. The system's data (some of which are classified), computational spreadsheets, and operating modules are installed in a central database and accessed by authorized users from their local computer workstations over a secure network. If you have access to a network where the system resides and enabling software has been installed on your local computer,[4] you may participate in the tutorial by following instructions that appear in boxes distributed throughout this manual.

CONTENTS OF THE MANUAL

Chapter Two sets up the hypothetical study used throughout this manual to illustrate FSC System features. The study is a general exercise to consider alternative cost-saving measures that might be motivated by a substantial reduction in Defense Fiscal Guidance. In that context, the analyst is likely to examine several alternative "cases," i.e., alternative scenarios with associated assumptions and methods, each with its own internally consistent cost results. Chapter Two lists the several study and case design options you can implement when using the FSC System, indicating which are illustrated in this manual.

The illustrated study specifically explores the effects of deactivating an entire Army division. Individual cases compare alternative assumptions about how reductions in equipment requirements might be implemented. In addition, the Appendix shows how to use an FSC "template"—a notional forces construct—to examine the additional effects of deactivating supporting units in Echelons Above Division (EAD).

Chapters Three through Eight (and the Appendix) explain how you can conduct various steps in the illustrative analysis. Each section describes a function available in the FSC System—viewing data, specifying a forces deactivation, computing equipment cost effects, and so on—and notes various options for using each function in alternative studies.

Each chapter also contains tutorial instructions and figures showing what a computer monitor would display as these instructions are carried out. To allow this manual to be unclassified, the figures use an unclassified demonstration database that covers fiscal years (FY) 1995–1999. However, the display formats are the same as those for the current, classified FSC database.

Finally, Chapter Nine contains brief concluding comments.

_____

[4]For information on gaining access to the FSC System, contact the authors of this manual at RAND, Santa Monica.

Chapter Two

SETTING UP A COST STUDY

This chapter lays a foundation for relating the tools and techniques available in the FSC System to common tasks undertaken in any force and support costing study. The chapter identifies the core study design decisions you, as an analyst, make when setting up a study, and indicates the options you can implement using the FSC System. The chapter also specifies a particular study design that will be used for illustrative purposes throughout this manual.

OVERVIEW OF THE ANALYSIS PROCESS

In any costing exercise, you normally undertake a number of subtasks: general planning and information-gathering, collecting and reviewing data, choosing assumptions and methods, doing calculations, and reviewing the results. Time permitting, you may iterate through some of these subtasks several times, exploring alternative "what-if?" scenarios and comparing their results.

The analysis process with the use of the FSC System is illustrated in Figure 2.1. It supports all of the subtasks described above, and they can be performed in almost any sequence, as the arrows suggest. The FSC System is structured to provide tools and databases to assist you in conducting your analysis as you normally do.

The figure labels the subtasks using standard FSC terminology. Most of the labels are self-explanatory, but a few have precise meanings in the FSC System, as follows:

- **Case:** A particular combination of assumptions, databases, and methods that together generate a single, internally consistent set of cost results. A number of alternative cases may be generated and compared as part of a study concerned with a given topic or issue.

- **Action:** A record of a particular set of changes ("deltas") in forces, assets, or infrastructure that are specified as part of a case. A case typically includes several actions, e.g., to record deltas in the number of Army units, corresponding deltas in Army enlisted and officer end-strengths, and corresponding deltas in equipment of various kinds.

- **Method:** A set of modeling techniques and assumptions applied to estimate cost effects for a given action. For example, a simple way to analyze a cut in enlisted

3

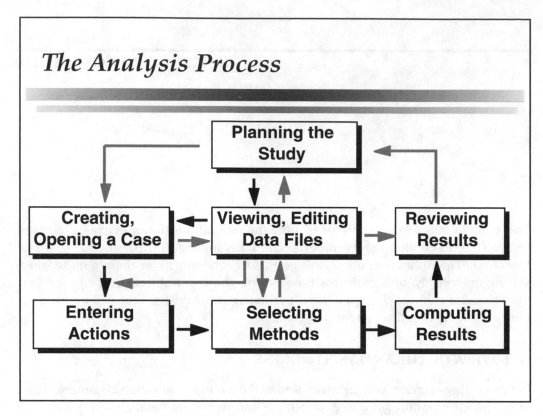

The Analysis Process

Figure 2.1 — Overview of the Analysis Process

personnel is to apply an average cost factor to the change in end-strength; an alternative method is to model the way average costs change over time as the grade and length-of-service profile of the enlisted force adjusts to a cut in accessions. The FSC System lets you choose between these methods, as well as impose additional conditions or constraints.

For illustrative purposes, this manual proceeds through the subtasks in the order suggested by the dark arrows in the figure; in the remainder of this chapter, you will lay out a general study plan. In subsequent chapters, you will examine some background data, begin your first case, enter the actions to be costed, select methods and compute results, and finally review results.

CORE DESIGN DECISIONS FOR THE STUDY PLAN

To provide a context for your study design decisions, let us suppose that your study is motivated by an expected cut in the Defense Fiscal Guidance. You have been assigned to suggest cost-cutting measures that could help bring the President's Budget and the POM period (the six future years covered by the defense Program Objective Memorandum) into alignment with the revised guidance.

Many cost-saving measures—and combinations of measures— might be considered. The FSC System is not designed to choose one for you. (You cannot simply enter a spending target and let the FSC System calculate how to meet it.) Instead, based on your own expertise and knowledge of the situation, you lay out an initial cost-cutting approach, then use the FSC System to explore its implications.

In principle, you could lay out a cost-cutting option in considerable detail before you begin using the FSC System. In practice, however, most study designs evolve during the course of the investigation, as new information leads to new ideas. Using the FSC System, you only need to make a few basic design decisions in order to begin your study. As summarized in Figure 2.2, they are:

- **The type of adjustment.** The FSC System is designed to analyze adjustments in force structure, in manpower or equipping policy, and in infrastructure management.[1] You can make changes in only one category and allow the FSC System to infer the effects in other categories, or you can control adjustments in all categories. For example, if you cut forces, you can allow the FSC System to infer that the reduction in manpower requirements will lead to a reduction in end-strength, or you can intervene to hold end-strength constant.

- **The affected organizations.** The FSC System lets you consider cuts (or additions) to one or more components (Active, Reserve, Guard) in one or more Services.

Basic Planning Considerations

- **Type of adjustment: <u>Forces</u>? Assets? Infrastructure? Combination?**
- **Affected organization(s):**
  - <u>Army</u>? Navy? Air Force? Marines? Joint?
  - <u>Active</u>? Reserve? Guard? All components?
- **Size of adjustment: Small and/or localized? <u>Large</u> and/or pervasive?**
- **Timing of adjustment: <u>POM period</u>? Long-term?**

Figure 2.2 — Options for Key Study Design Decisions

---

[1]At this writing, the FSC infrastructure modules remain to be developed. However, some changes in infrastructure management (e.g., consolidation of functions) can be approximated by editing the relevant cost factors.

Similarly, you can cut (or expand) manpower or equipment in alternative Services.

- **The size of the adjustment.** The FSC System can analyze small or large changes in force or support structure. For example, an Army study could deactivate (or activate) only a single company or battalion or as much as several divisions.

- **Time frame of the adjustment.** The default time frame for a case analysis in the FSC System is the current year through the end of the POM period. However, you can reduce the period to as few as three years or extend the analysis up to 30 years into the future.

The context of your hypothetical study suggests that you will want to consider major cuts in the near future, and you will be especially interested in cost effects during the POM period. Let us also suppose you plan to focus on a major cut in the Army active component because you expect that action to generate sizable downstream savings, as a result of reductions in asset and infrastructure requirements and associated operations. Thus, as the underlined options in Figure 2.2 indicate, you intend to cut Army active forces substantially and trace the effects through the end of the POM period.

The choices you have made so far do not constitute a full study design, but they establish a basic outline sufficient to get started using the FSC System. In the next chapter, you will see how using the system's databases can help you refine your study plan and move toward specifying a particular case.

Chapter Three

STARTING THE FSC SYSTEM AND VIEWING ITS DATABASES

This chapter begins your hypothetical study by starting up the FSC System and using its menu bar and dialog boxes to access data describing the structure and composition of Army active forces. The objective is to select a portion of the Army active force structure to use in your study of the potential cost savings from a major deactivation.

To operate the FSC System, you must have network access to the FSC central database, and the FSC System startup files must be installed on your local computer workstation. If these conditions are met, you can begin the FSC tutorial by following the instructions in boxes like the one below. Otherwise, refer to the figures in this section to see what your computer monitor would display as you begin using the FSC System.

> **TUTORIAL:** Deliveries of the system are accompanied by start-up instructions specific to your computer platform (PC or Macintosh). Follow the start-up instructions now. After the system verifies that all directories and network connections are in order, it displays a "Welcome" screen and menu bar.

FSC OPENING DISPLAY AND MENU BAR

When you start the FSC System, it displays the "Welcome" screen and menu bar illustrated in Figure 3.1. The system version illustrated here is labeled "95a" because this manual uses an unclassified, demonstration database that covers the period 1995–1999. Actual FSC installations display a different version label.

The menu bar along the top of the screen offers drop-down menus that enable you to initiate various activities, as follows:

Casefiles: Start a new cost exercise (called a "case"), or open files from a previous one.

Factors: Examine and/or edit variables, such as pay rates or costs per flying-hour, that are used in cost calculations.

Programs: View tables showing forces and assets currently programmed in the defense budget and POM.

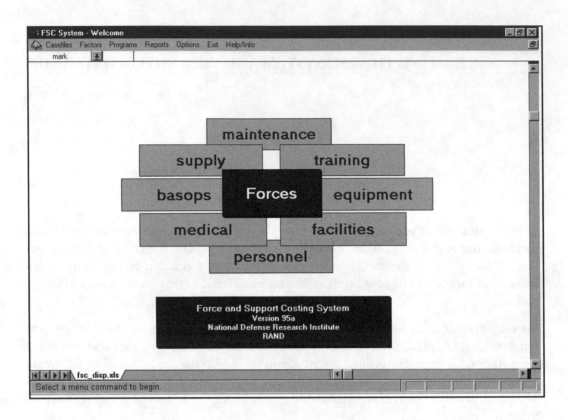

Figure 3.1 — FSC System Welcome Display and Menu Bar

Reports: View detailed output reports from previous FSC cases.

Options: Use an FSC utility, e.g., to add a note to a worksheet cell.

Exit: Display the standard Excel menu bar or quit Excel (and FSC) entirely.

Help/Info: View online help and documentation for the FSC System.

Notice that a message at the lower left of the screen (the first of many such messages that will appear as you use the FSC System) advises you to "Select a menu command to begin."

TUTORIAL: As the FSC System expands, additional menu choices are added to the standard menus. If you are not using FSC demonstration version 95a, for which menu choices are listed in this manual, you may wish to familiarize yourself with your version of the FSC System by clicking on each menu in turn and reading its contents.

USING MENUS TO VIEW DATA

Data are accessible from the **Factors** and **Programs** menus. The choices under those two menus in the demonstration version of the FSC System are listed in Table 3.1.

The menu commands are similar under the two menus, but they access different kinds of data.

Factors	Programs
Army forces factors...	Army structure...
USAF forces factors...	USAF structure...
Personnel...	Personnel...
Army equipment...	Army equipment...
USAF equipment...	USAF equipment...

Table 3.1

Factor and Program Menu Contents in FSC System Version 95a

The **Factors** menu accesses data that portray ratios, rates, and other standardized measures that are commonly used in cost exercises. For example, Army forces factors include the average share of the total Army Reserve enlisted force that is on full-time active duty. Forces factors also portray notional force structures, such as a typical armored division or its notional complement of supporting forces in echelons above division. Notional force structures are called "templates" in the FSC System, and you can learn about accessing and using them in the Appendix.

Notably, you can customize the factors database by editing the values the FSC System provides and saving the edited files at your local workstation. To do that, you (a) choose the appropriate factors category from the **Factors** menu, (b) fill in dialog box options to select the particular factors file of interest, (c) view and edit entries in that file, and (d) use the "Save..." button on the file to save it as a new, customized file at your local workstation. Later, when you do a cost analysis, the FSC System will let you choose between using factors from the FSC central database (the default choice) or your own edited values.

In contrast, the **Programs** menu leads to data describing the forces and assets actually programmed in the current defense Budget and POM. These data cannot be customized because doing so could create substantial inconsistencies among forces, personnel strengths, equipment inventories, and other program variables that are interrelated in the budget and POM.[1]

For your hypothetical study, assume that you want to view data describing the programmed Army active force structure. Using the **Programs** menu, you can view Army forces and decide what portion of them to use in your analysis.

VIEWING PROGRAM DATA

The FSC Programs database contains detailed information on the entire force structure programmed to appear in a Service in each fiscal year from now to the end of the

---

[1]Program data on units, manpower, and equipment inventories are developed by different organizations at slightly different times. Database preparation for the FSC System includes cross-check and reconciliation steps that provide an internally consistent portrayal of the overall force and asset structure.

current POM period. Combat units, such as Army battalions or Air Force aircraft squadrons, are grouped by type of unit, location, component, and other characteristics; you can view either combat unit counts (i.e., numbers of like units) or estimates of the manpower and equipment to be found in those units. Supporting force organizations are measured by quantities of personnel and equipment assets, grouped into supporting force categories such as logistics or security.

Despite these groupings, the program database for each Service is very large—so large that it cannot be displayed in its entirety in a single Excel spreadsheet. Therefore, when you pull down the FSC Program menu and select "Army force structure," the FSC System will ask you to select a portion of the data to be displayed. The dialog box for that selection appears in Figure 3.2.

For your hypothetical study, you are interested in identifying an active division that is likely to be a source of considerable costs. An active armor division might be a good candidate because such divisions are large in terms of personnel strength, and generally contain equipment that is relatively costly to acquire, maintain, and operate. In practice, however, armor divisions are in various stages of activation and may not be fully manned and operational. Therefore, to select a fully operational armor division, you might examine the manpower strengths of various programmed active Army armor divisions. As Figure 3.2 illustrates, you can do that by selecting options to view manpower counts for all Army Armor divisions.

> **TUTORIAL:** From the FSC "Welcome" screen, pull down the **Programs** menu and select "Army force structure..." In the dialog box, choose "Troop Program" (the active selection when the dialog box opens). Then select "Division" and "Armor" for Category and Branch but leave "Unit Number" at ALL so that all such divisions will be displayed. Then select "Manpower counts" as the display option. When these selections are completed, click the "Do Extract" button.

Figure 3.3 shows the program data display for the selections made in Figure 3.2. The FSC System displays data in Excel worksheets that have special FSC header sections. The header identifies the data being shown and contains several buttons that allow you to take various actions, such as saving a copy of the worksheet, printing it, or "canceling" it to return to a previous FSC display.

The demonstration FSC database used for this manual contains only one active armor division that is fully manned with active-component personnel throughout the POM period—the 998th Armor Division. That division will be used in the next chapter to begin the cost analysis. If you are using a different FSC version, you would select a different division to use in your first case.

> **TUTORIAL:** Click the "Cancel and return" button on the pivot table worksheet. If you wish to break from using the FSC System and return to it later, use the Exit menu and select "To windows" ("To finder" on a Macintosh). This will shut down the FSC System and Excel. Return to the tutorial by restarting as you did at the beginning of this chapter.

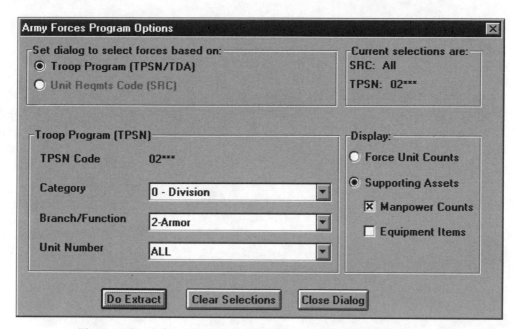

Figure 3.2 — Selection Dialog Box for Programs: Army Force Structure

FSC System

Casefiles Factors Programs Reports Options Exit Help/Info

A6

Army Program Listing

CANCEL and return

ARMY PROGRAM DATA DISPLAY
Manpower for SRC/TPSN: ALL/02

Save   Print

Assets

ORG	UNIT	BRN	TPSN	COMPO	REGIO	TYPE	ASSET	ASSET_NAME	Data FY94	FY95	FY96	FY97	FY98	FY99	FY00
							WOF	WARRANT	0	0	0	0	0	0	
	997th	ARM	2997	ACT	CON	PERS	CIV	CIVILIAN	0	0	0	0	0	0	
							ENL	ENLISTED	12,643	12,569	12,834	12,860	12,860	12,860	12,89
							OFF	OFFICER	1,137	1,119	1,143	1,151	1,151	1,151	1,14
							WOF	WARRANT	302	321	328	321	321	321	32
					EUR	PERS	CIV	CIVILIAN	0	0	0	0	0	0	
							ENL	ENLISTED	0	0	0	0	0	0	
							OFF	OFFICER	0	0	0	0	0	0	
							WOF	WARRANT	0	0	0	0	0	0	
				NG	CON	PERS	CIV	CIVILIAN	0	0	0	0	0	0	
							ENL	ENLISTED	2,416	2,955	3,191	3,283	3,283	3,283	3,28
							OFF	OFFICER	150	191	201	238	238	238	23
							WOF	WARRANT	11	12	16	19	19	19	1
	998th	ARM	2998	ACT	CON	PERS	CIV	CIVILIAN	0	0	0	0	0	0	
							ENL	ENLISTED	2,567	2,646	2,646	2,646	2,646	2,646	2,64
							OFF	OFFICER	204	198	198	198	198	198	19
							WOF	WARRANT	9	10	10	10	10	10	1
					EUR	PERS	CIV	CIVILIAN	0	0	0	0	0	0	
							ENL	ENLISTED	14,716	14,719	14,625	14,625	14,625	14,675	14,67
							OFF	OFFICER	1,136	1,139	1,145	1,145	1,145	1,147	1,14
							WOF	WARRANT	292	251	247	247	247	249	24
Grand Total									87,553	90,518	90,721	89,937	89,950	90,513	90,54

Ready

Figure 3.3 — Program Data Pivot Table Display

Chapter Four

BEGINNING A NEW CASE

In this chapter, you will begin your first FSC case, i.e., specific cost analysis. You will create an FSC worksheet—called a "casesheet"—that will record the actions for the case and summarize the cost findings. This chapter illustrates and describes the general format and contents of a case worksheet and shows you how to create and initialize one for a new case.

BASIC INFORMATION ABOUT CASEFILES AND CASESHEETS

The FSC System records a lot of information about each case you analyze. Separate worksheets record the changes you make in forces, assets, or infrastructure, and other worksheets report the detailed cost effects. Taken together, all of the worksheets associated with a given cost exercise are known as a "casefile." You initiate, save, and open casefiles by using the **Casefiles** menu; its commands are listed in Table 4.1.

Table 4.1

Casefiles Menu Contents

Casefiles
New
Open
Close
Save as ...
Delete

NOTE: The delete option for the Casefiles menu has not been implemented at this writing.

The principal worksheet in a casefile is called the "casesheet" (short for "Case Specification Worksheet"). You begin a new cost exercise by creating a new casesheet, and you use buttons located on it to control the development of the case and compute its findings. The casesheet in turn records the various steps you carry out during your analysis and summarizes the results.

You do not have to keep track of all of a casefile's worksheet names and locations. The FSC System does that for you. It lists supporting the files on the casesheet and stores them in special subdirectories at your local workstation. Moreover, it regularly saves your FSC worksheets as you work, so you do not need to remember to save as you go.

Keep in mind, however, that the FSC System can only perform these functions if you do not independently move or rename any of your FSC worksheets. Always use the "Save as..." command under the FSC **Casefiles** menu when you want to resave a casefile with a different name or in a different directory. Also use that command if you want to create a backup copy of your casefile.

CREATING AND INITIALIZING A NEW CASE

When you select "New" from the **Casefiles** menu, the FSC System displays a raw casesheet and a dialog box asking you to provide a case name and choose the time horizon the case will analyze. You must enter up to eight characters for the case name; for your hypothetical exercise, let "cut998" be the case name. The FSC version you use automatically determines the starting year for the analysis (e.g., the first analysis year in demonstration FSC Version 95a is 1995), and the dialog box will offer the end of the corresponding POM as a default for the case time horizon. We will use the default for the illustrated analysis, but you could enter a different year to shorten the time frame to as few as three years or extend it as far as 30 years. Figure 4.1 shows the dialog entries for your hypothetical case.

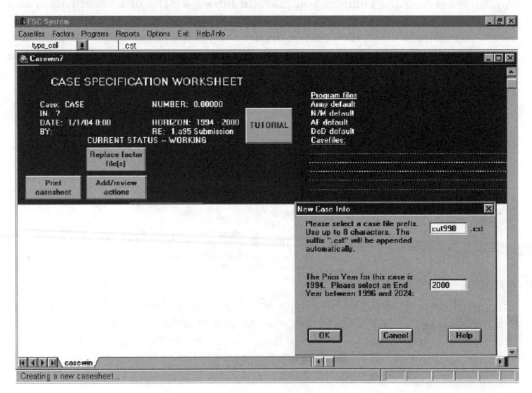

Figure 4.1 — Naming and Initializing a New Casesheet

Another dialog box (not shown) asks where you want to store your casefile. All case-files are automatically stored in a special FSC directory (*fsc_user*) on your computer's local drive. You can save your new casefile directly in that directory or create a new subdirectory for this case. After you make your selection in the second dialog box, the FSC System initializes and displays your newly established casesheet.

> TUTORIAL: Go to the FSC Welcome screen. (If necessary, start up the FSC System. Alternatively, if any FSC worksheet is open, click the "Cancel and return" button.) Pull down the **Casefiles** menu and select "New." Enter a case name in the first dialog box. (Use up to 8 characters and follow other instructions in the dialog box.) Then click OK. In the next dialog box, create a subdirectory if desired, or simply click OK. The FSC System will then fill in some information in the new casesheet header area and display the full casesheet.

CASESHEET CONTENTS

Figure 4.2 shows the monitor display after a new casesheet is initialized. The FSC menu bar is still present and can be used to view data or reports or access the FSC Help/Info window while the casesheet is open. On the casesheet itself, the blue header area (shown here as black) records information about the case, such as the date it was created and a unique case number that will also appear on detailed output reports in this casefile.

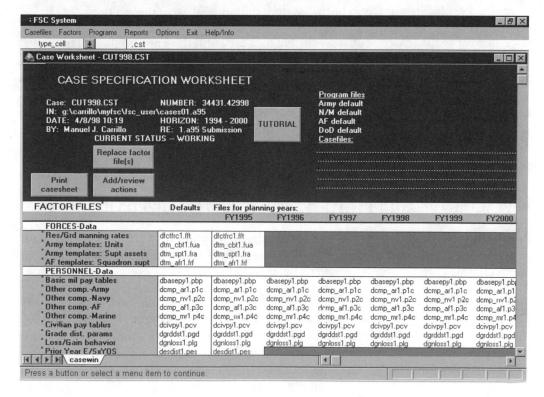

Figure 4.2 — A Newly Initialized Casesheet

The header area also displays buttons to operate currently available casesheet procedures; as you carry out your analysis, buttons that are no longer relevant become hidden and others become visible.

There are more sections in the casesheet than Figure 4.2 reveals. By scrolling down the casesheet, you would observe that it contains the following sections:

- **Header** area for basic case information and control buttons.

- **Factor Files** area listing the factors data files used on the case. A new case lists central database files by default, but customized factor files can be substituted as described in Chapter Three.

- **Actions List** to record actions as they are entered during the analysis. Action records are grouped by type, and report the action status (i.e., whether each one has been computed or not), the files where the action input and output data are stored, the cost method used, and other conditions or constraints applied to the actions.

- **Case Modification Log** to note whether any actions are deleted or revised after they have been computed. If you delete or revise a computed action, its former effects are completely removed from the current case, but its output files remain available.

- **Delta Cost** section for a summary of cost results by action and major appropriation. Results are shown in both constant dollars and escalated values.

On a newly created casesheet, only the Header area and the Factor Files section contain information. Other sections are filled in when you specify the case and compute costs. (You will see entries in the Actions List and Delta Cost sections later in this manual.) Because the casesheet is too large to view all at once, the system automatically displays whichever section is relevant and hides other sections from view when you click a casesheet button. Whenever some portions are hidden, the casesheet will display a "Show full casesheet" button in the header area; you can click it to see and scroll through the entire sheet.

REPLACING DEFAULT FACTOR FILES WITH CUSTOMIZED VERSIONS

Recall from Chapter Three that the FSC System lets you edit factors and save your own custom versions of the factor files. If you have done so, you can use the custom factors in this case.

To do so, scroll to the top of the casesheet (or click the "Show full casesheet" button). Then select the casesheet cell that names the FSC default version of the file, and click the "Replace factors file" button; a dialog box opens to let you choose one of your edited files, and the FSC System enters its name on the casesheet. Some factors files contain factors for a single year, and those files are listed in separate columns on the casesheet. You can substitute a customized factor file for the default file in as many years as you wish. For example, you can apply default basic pay rates to the first two years of the analysis, but switch to a customized basic pay table for all the remaining years.

The hypothetical case illustrated in this manual relies on the standard FSC factors. You are welcome to use custom factors; just remember to do so before you compute actions in which the factors are used. When actions are computed, the Factors File section becomes locked and the file replacement option is no longer available.

RETRIEVING AND REVISING CASES

Although this chapter, illustrated the creation of an entirely new case, it is also possible to form a new case by retrieving and revising a previous one. To do that, you simply open a previously completed case and resave it with a new case name (using "Save as..." under the **Casefiles** menu). Then you can add actions to the case (see Chapter Five) or revise the actions already on the case. The Case Modification Log keeps track of any actions you revise so that you can still locate the results they produced. A case constructed by modifying an existing case is illustrated in the Appendix.

ENTERING A FORCES ACTION

In this chapter, you will construct a forces action that deactivates an entire Army division. Activations and deactivations of forces generate cost effects by affecting operations, assets, and infrastructure—and that is how the FSC System models them. Thus, you will create a forces action by entering deltas in Army forces. Then you will "compute" the forces action in order to generate secondary actions consisting of deltas in personnel, equipment, and infrastructure. Later sections will then show how to compute the secondary actions to produce cost results.

INITIATING A CASE ACTION

So far, you have decided that your case will include one central Action—one that cuts all units in the 998th Armor division from the Army program. To begin constructing that action, you click the "Add/review actions" button on the casesheet. It displays the upper dialog box shown in Figure 5.1. The action you have in mind involves entering deltas for *Forces* in the *Army*, so those are the options selected in the illustrated dialog box. You can see, however, that the dialog box offers you options for adding other types of actions to your casesheet, such as actions that would modify personnel end-strengths or equipment inventories in any Service. Clicking the "Add/review actions" button is always the first step in creating any type of action for your case.

Also use the "Add/review actions" button whenever you want to review an action that was created previously for your case, or to "borrow" one from a previous case. The details of every action you create are recorded on a separate action worksheet, and the "Add/review actions" button gives you access to those worksheets. When you make selections in the dialog box, the FSC System checks whether any action worksheets of the proper type already exist. If so, another dialog box (such as the one shown below in Figure 5.2) will offer the option of reviewing those actions.

CONSTRUCTING A FORCES EXTRACT

To enter deltas in forces, you first need to access data on the currently programmed force structure. (The FSC System is designed to make sure that the actions you build are feasible within the programmed force structure, e.g., to avoid estimating the

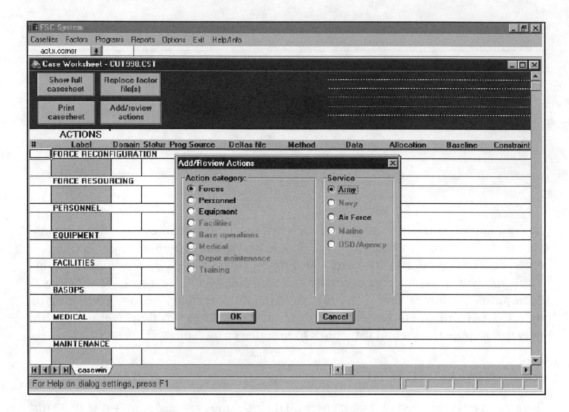

Figure 5.1 — Dialog Box to Add/Review a Case Action

savings from deactivating units that do not exist.) When you used the **Programs** menu to view Army forces data in Chapter Three, you learned that the Army forces database is too large to display on a single worksheet, and you made dialog box selections to choose a portion of the forces to be displayed. The same happens when you want to enter a forces action; the FSC System asks you to select a portion of the force structure—called an "extract"—to work on.

However, there is an important difference between selecting forces data from the **Programs** menu and constructing a forces extract. In Chapter Two, you could choose to see either counts of forces units or the assets assigned those same units. In contrast, an action extract measures each portion of the force structure in a particular way that enables the data to be used for cost analysis. For that purpose, "forces" are divided into three measurement categories:

- **"Direct" combat forces.** These are combat units that have standardized personnel and equipment requirements specified in the FSC database. The Army identifies them by Standard Requirements Code (SRC), and the FSC System stores asset requirements on roughly 2,000 such units.[1] The FSC System also

---

[1]There are over 4,000 distinct Army SRCs for battalion- and company-level units. Many of those are for units that are rarely observed in Army programmed forces, or represent such minor variations in asset

identifies common asset standards for combat units in other services, such as typical flying squadrons in the Air Force. Because these units are standardized, the FSC System can represent them by counts of like units.

- **Additional assets assigned to direct combat forces.** In the Army, these additional assets are labeled "TDA (Table of Distribution and Allowances) Augmentation." They mostly consist of additional manpower assigned to combat units during peacetime, though additional equipment may also be assigned. These assignments are not standardized across units, so the FSC System reports them directly in terms of asset quantities assigned to particular divisions or separate brigades.

- **Assets assigned to forces other than direct combat units.** In the Army, these forces are found in Nondivisional Combat (NDC) or Combat Service Support (CSS) units in echelons above division (i.e., assigned to Corps or Armies). These units vary too widely in size and structure to be measured by counts of like units, and they are not assigned to individual divisions or separate brigades. (Similar supporting force structures exist in other Services as well.) The FSC System measures these Army forces by asset quantities within NDC or CSS branches, such as Engineers or Composite Services.

Categories of forces must be displayed in different action worksheets because they are measured differently. If you want to modify forces in more than one category on a given case, you must create a separate action for each category. In this illustration, your action will involve only direct combat forces, but the Appendix shows how to add an additional action to modify forces in echelons above division as well.

After you select Army forces in the "Add/review actions" dialog, the next dialog box asks whether you wish to extract a new set of program data or see an extract you created earlier, perhaps during a previous case. (This dialog box also offers the option to review an existing extract that was previously constructed for this or another case.) You will choose a new extract for your hypothetical case, so the next dialog will ask you to specify the portion of forces to be displayed. These two dialog boxes for forces extract selection are overlaid in Figure 5.2.

When the second dialog box opens, the section labeled "Forces measured by..." is already set to select direct combat unit counts. You can change that to show asset quantities if you wish to extract TDA Augmentation or NDC/CSS data, but our illustration is for an extract of unit counts.

The upper left section of the extract dialog box allows you to select a portion of the force structure based on any of four criteria:

- **Troop Program Sequence Number (TPSN)** enables selection based on organizational identity. This option allows you to select unit counts for particular divi-

---

requirements that they do not generate noticeable differences in costs. The FSC database has been pared down to a meaningful subset of SRCs by collapsing similar units into representative forms.

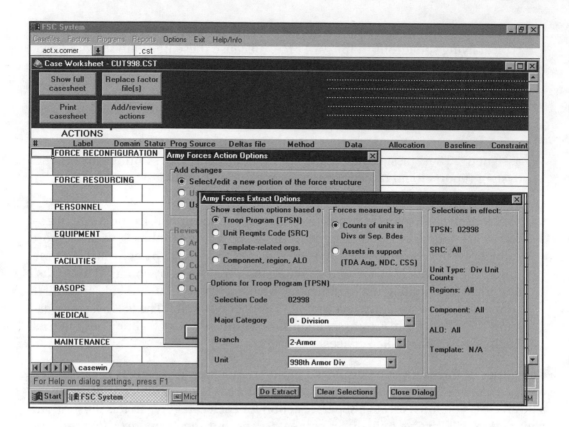

Figure 5.2 — Dialogs to Select a New Force Extract

sions or separate brigades. Alternatively, you could select asset quantities for TDA Augmentation in those same divisions or separate brigades, or for NDC/CSS branches.

- **Unit Requirements Code (SRC)** enables selection of all battalion/company level combat units by type, without regard to the divisions or separate brigades to which they are assigned. This criterion can only be used to extract unit counts, not unit assets.

- **Template-related organizations** enables selection of forces in combat units or NDC/CSS branches that match the elements of notional force structures, such as a notional armor division or the notional forces in EAD that support a particular type of division or separate brigade. (See Appendix for an illustration using this criterion.)

- **Component, region, or ALO** enables selection of direct combat or supporting forces according to their component (active, Reserve, Guard), their peacetime location (CONUS, OCONUS, or Alaska/Hawaii),[2] or their Authorized Level of

---

[2]The FSC program database records forces locations in eight world regions. However, only three geographic areas are used for cost estimation because available data do not permit closer distinctions. Personnel cost rates differ between "CONUS(+)" (the continental U.S. plus Alaska and Hawaii) and the rest

Organization (an indicator of the extent to which their wartime asset requirements are filled during peacetime).

For the exercise illustrated in this manual, you select forces by TPSN, and you choose the 998th Armor division in the drop-down dialog lists. When your selection is complete, you click "Do extract" to tell the FSC System to query the central database and display the forces you selected.

> **TUTORIAL:** If necessary, follow the tutorial instructions in previous sections until you have a new casesheet displayed on your monitor. Click on the "Add/review actions" button in the casesheet header area, select "Army" and "Forces" in the dialog, and click OK. In the Army Forces Action Options dialog box, select "Select/edit a new portion of the force structure" and click OK. In the Army Forces Extract Options dialog box, make sure the option for measuring forces by unit counts and the TPSN criterion option are selected, and choose the division you intend to analyze in the drop-down lists. Finally, click the "Do Extract" button. The FSC System will query the central database and, after a few seconds, will display your Army forces action worksheet.

SPECIFYING FORCE DELTAS ON AN ACTION WORKSHEET

All FSC action worksheets are readily identifiable by their red header areas. A title box in the header area indicates what type of action worksheet you are viewing, and buttons in the header area control processing as you enter the force, asset, or infrastructure deltas for the action. Figure 5.3 illustrates the newly created Army forces extract you specified for the first action on your hypothetical case. It lists currently programmed unit counts by component, ALO, region, and fiscal year and for each of the types of units in the Army 998th Armor division. The figure also shows one of the dialog boxes (discussed below) that is triggered by clicking a button on the worksheet.

You specify the forces deltas to be analyzed on your case by editing the data reported for the currently programmed force structure. You can edit any of the unit counts shown onscreen in blue font (columns for FY 95 and beyond here). You can edit the counts in each individual year for each individual unit, or you can use worksheet buttons to make more comprehensive changes. For example, the "Add a unit" button lets you insert a new data row to represent another type of unit, and you can enter new counts for that unit. Or you can edit a unit count in a particular year and use the "Fill outyears" button to extend the edited value to all the subsequent years.

The easiest way to deactivate the entire division is to use the "Change All" button. It produces the dialog box shown above in Figure 5.3, which allows you to change all the unit counts to zero (the *Deactivate all* option) or to multiply all of them by some

---

of the world ("OCONUS(–)"). Equipment cost rates, which include the cost of moving equipment between units and maintenance and supply depots, differ between the continental U.S. ("CONUS(–)") and the rest of the world including Alaska and Hawaii ("OCONUS(+)"). Thus, forces extracts only need to distinguish among CONUS, OCONUS, and AK/HI locations.

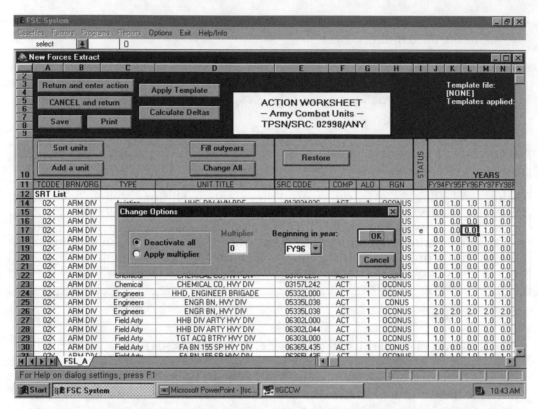

Figure 5.3 — Army Forces Action Worksheet with "Change All" Dialog Box

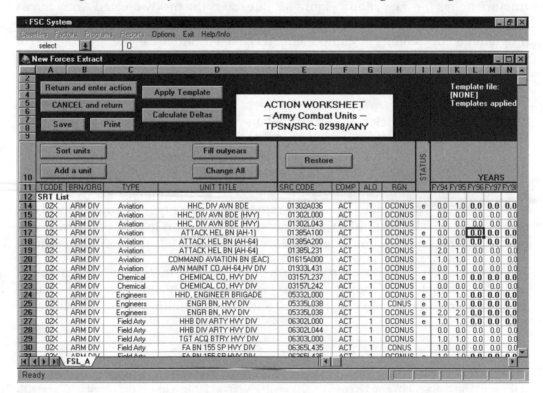

Figure 5.4 — Army Forces Edited by Applying the "Change All" Button

number (the *Apply Multiplier* option) beginning in any fiscal year. For example, if you want to deactivate the division over two years, you could multiply all the unit counts in some year by 0.5 and then deactivate them all in the following year.[3]

For this illustration, however, we assume you deactivate the entire division in FY96. When you click OK in the dialog box, the revised unit counts are entered in the worksheet in bold to make them readily apparent, as illustrated in Figure 5.4. In addition, the notation "e" is entered in the status column of each affected row. This can help you distinguish between units that were already programmed for deactivation (such as the Attack Helicopter Battalion shown in the sixth row of data) and those that you specifically deactivated. Only the deactivations that you specify will generate cost effects on your case.[4]

To cost out force actions, the FSC System must first convert the edited forces data into deltas—numbers of units added or subtracted in each fiscal year. On the action worksheet, you can click the "Calculate deltas" button to see the deltas corresponding to the entries you made. Alternatively, you can simply click the "Return and enter action" button. It also calculates the deltas, but does not stop to display them for you. Instead, the system simply saves and closes this file, and records the new action on the casesheet.

TUTORIAL: Decide which year you would like the deactivation to begin. If you select any unit count in the worksheet column for that year, the "Change All" dialog box will open with that year already selected. To open the dialog box, click the "Change All" button. When the dialog box opens, select the *Deactivate All* option if it is not already selected, and check to make sure the proper beginning year appears in the drop-down list. Then click OK. After the FSC System finishes editing the unit counts, you may wish to scroll through the worksheet to view and perhaps add additional edits to the unit counts. (If desired, you can now click the "Calculate Deltas" button and, when the FSC System finishes calculating, scroll through the list of forces deltas to review them.) Finally, click the "Return and enter action" button to close this worksheet and enter a record of it on your casesheet.

RECORDING FORCES ACTIONS ON THE CASESHEET

You can add many actions to a given case. For example, after entering the action that deactivates the 998th Armor division, you might add another forces action that cuts all the assets in TDA Augmentation for that same division. Or you might add an ac-

---

[3]The FSC System calculates costs by assuming that assets and operations adjust linearly between the beginning and ending of a fiscal year. For example, if a unit is deactived during a fiscal year, its operations are assumed to wind down linearly to zero during the year, which implies that total operations for the year will be half of their normal level. Accordingly, deactivating "half" of a unit corresponds to reducing the year's total operations by one quarter, leaving half of the unit's personnel and equipment intact at year end. To do this, the FSC System permits asset quantities to be measured in fractions. This produces good cost approximations in most cases, but users may need to adjust procurement cost estimates for actions that involve very small numbers of very costly equipment items.

[4]Many cost models estimate savings from deactivations for notional divisions without regard to how actual divisions would be configured during the Budget and POM period. The FSC estimates of savings from deactivating an actual division may differ from estimates by other cost models, even when they apply the same cost factors, because actual divisions rarely match notional ones exactly.

tion that adds or deletes forces in another Service. As you construct actions, the casesheet automatically keeps a record of what you have done. Each action record begins with a three-character key code and a descriptive label, and includes the name of the action worksheet on which you specified the deltas to be analyzed.

Figure 5.5 shows the casesheet action section as it appears just after you deactivate the 998th Army division. Although you created only one forces action, two action records have been added to the casesheet. The first one is listed under the heading "Forces Reconfiguration." It begins with the key code "x01" and shows the name of the action worksheet (CUT998.X01, but in lower case to save space) in the Deltas file column. If you add more forces actions, there will be additional records in the Force Reconfiguration section, and their key codes would be "x02," "x03," and so on.

The other action record in Figure 5.5 begins with the key code "faf" and appears under the heading "Force Resourcing." To make FSC processing more efficient—and to make sure that the actions you enter do not cut the same forces more than once[5]— the system automatically keeps a running total of all the deltas you enter. All deltas

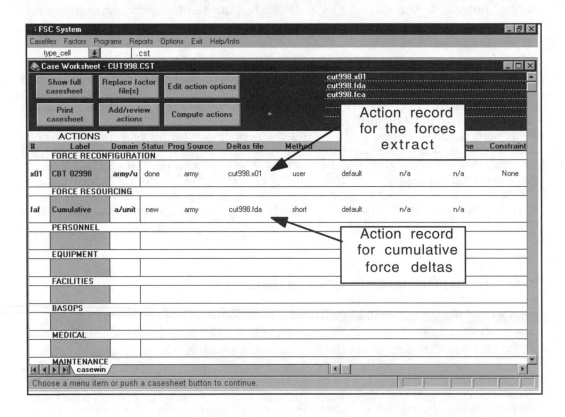

Figure 5.5 — Casesheet with Newly Entered Forces Action

---

[5]The FSC System compares the cumulative deltas against currently programmed forces when you enter a forces action. If your action attempts to cut forces in excess of the total program, you would receive a warning and an opportunity to modify the deltas.

in Army unit counts are in a worksheet named CUT998.FDA, as noted in the "faf" action record. (Its contents are illustrated in Chapter 6.) If you also specified deltas in forces measured by asset quantities, they would be summarized in another cumulative file and noted in another action record under the Force Resourcing heading, which is not illustrated here.

THE STATUS OF AN ACTION

The fourth cell in an action record indicates its status. "New" means that the action is awaiting analysis, whereas "done" means the action has been computed. Force Reconfiguration actions are always marked "done" because deltas are added to the cumulative totals as soon as an action is entered. However, the system does not analyze the cumulative deltas until you elect to compute them. That is why, in Figure 5.5, the status for the "faf" action is marked "new." Chapter Six tells how to compute this forces action.

COMPUTING A FORCES ACTION

In this chapter, you will compute the Army forces action that deactivates the 998th Armor division. As you will see, FSC forces models do not calculate costs directly. Instead, they translate forces deltas into deltas in assets and operations, permitting analysis of those effects and the costs they generate as distinct actions. This section explains how that is done, and examines the casesheet entries for the additional actions.

CONTENTS OF A CUMULATIVE DELTAS WORKSHEET FOR FORCES

To understand how forces actions are computed, it is helpful to examine a cumulative forces deltas file. The file generated in Chapter Five is shown in Figure 6.1. Each row shows the yearly deltas in the numbers of units that share the same Standard Requirements Code (SRC), component, location, and ALO. All of the deltas are less than or equal to zero because the only forces action on this case is one that deactivates Army forces.

This worksheet is for information only, and cannot be edited directly. To modify the forces deltas, you would go back and edit the original forces action worksheet or add a new forces action to the case, as Chapter Five described.

The FSC system also includes a worksheet that describes the forces remaining after applying your deltas. Army units that share the same SRC, component, region, and ALO are often found in several different divisions or separate brigades. Therefore, deactivating a single division probably does not mean deactivating all the units that share those characteristics. If you want to see how many such units would remain in the Army force structure, you can do so by clicking the "View corresponding force counts" button on the worksheet illustrated in Figure 6.1. You can return to the casesheet by clicking "Cancel."

COMPUTING A SINGLE FORCES ACTION

The process of computing a forces action involves translating forces deltas into deltas in assets and operations. When the deltas are measured by unit counts, as in Figure 6.1, the FSC System must determine the asset quantities for those units. When the

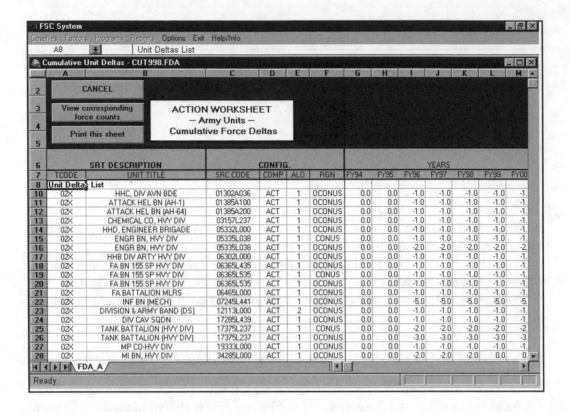

Figure 6.1 — Cumulative Deltas Worksheet for Army Direct Combat Forces

deltas are measured directly by asset quantities, the procedure involves transferring the asset deltas to the appropriate asset worksheets that enable cost effects to be computed.

Whenever at least one casesheet action has the status "New," a "Compute actions" button becomes visible in the casesheet header area. (See Figure 5.5 in the preceding chapter.) Before using the button, decide whether you want to compute a single action or have the FSC System continue processing—in "batch mode"—until every casesheet action is done. The choice matters, because computing a forces action causes additional actions to be added to the casesheet, and batch mode processing continues until all of those actions have been analyzed. Here, we will compute only a single Army forces action, and then pause to examine the newly generated asset actions.

> **TUTORIAL:** If you wish to view the cumulative deltas for your forces action, click the "Add/reviews actions" button on the casesheet, then select Army forces, and then choose the "View cumulative deltas" option in the next dialog box. After viewing the deltas, return to the casesheet by clicking the CANCEL button on the cumulative deltas worksheet. To compute the forces action, select any cell in the "faf" action record, then press the "Compute actions" button. In the next dialog, be sure the "Compute selected action only" option is selected, then click OK.

The FSC System computes the action by reading the "faf" file (without displaying it), and translating it into a database query. Heuristically, the query asks, "How much personnel strength and equipment inventory would be released by these units if they were removed from the Army force structure?" (Alternatively, of course, the query would estimate increases in manpower and equipment for units added to the force structure.) The FSC database contains data on the personnel and equipment requirements for each type of unit, as well as typical unit fill rates (i.e., fractions of asset requirements that are currently programmed to be filled in each year). The query does not estimate costs or savings at this point, but simply estimates the deltas in manpower and equipment associated with the specified unit deltas.

The query estimates deltas separately for several categories of personnel and many items of equipment. The deltas are stored in asset action worksheets, and asset actions are recorded on the casesheet. Figure 6.2 shows that computing a forces action that deactivates an entire division can produce a substantial number of personnel and equipment actions.

ASSET ACTION DOMAINS AND WORKSHEETS

On the casesheet, an asset action represents all the deltas within a category called a "domain." Each military personnel domain represents a combination of Service;

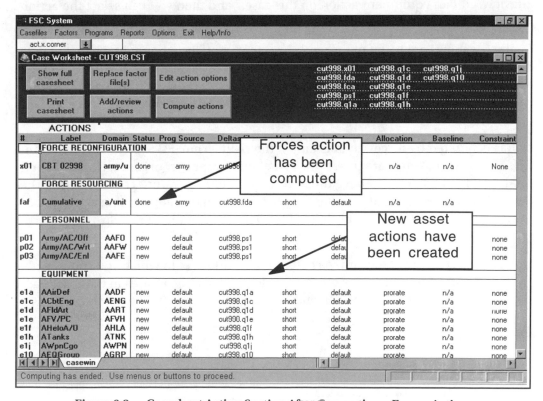

Figure 6.2 — Casesheet Action Section After Computing a Forces Action

component; full-time versus drill status; and officer versus warrant officer versus enlisted grade-range.[1]   There is also a separate domain for civilians in each combination of Service and component.[2]   A single personnel action worksheet collects deltas for all the domains in a given Service, which is why all the personnel actions in Figure 6.2 refer to the same deltas file. However, the casesheet lists a separate action for each personnel domain because that provides more flexibility in the choice of analysis methods or factors when the actions are computed.

Because the FSC System captures data on many equipment items, it would be burdensome to use a separate domain and compute a separate action for each item. Instead, each equipment domain contains several related items of equipment, such as types of Army field artillery. A given equipment action worksheet records deltas for all the equipment items within a single domain.

REVIEWING/REVISING ASSET ACTIONS DERIVED FROM FORCES ACTIONS

Recall from Chapter 5 that you can review any action on a case by pressing the casesheet "Add/review actions" button. Suppose, for example, that you are interested in the deltas for Army attack helicopters; click the "Add/review actions" button and choose *Equipment* and *Army* in the dialog box. The next dialog box lets you "Review/edit an equipment action on this case," and then you can select the action worksheet for attack helicopters.

Figure 6.3 shows a portion of the worksheet that contains a horizontal panel of data for each of several different helicopters.[3] The first few columns contain equipment identifiers, fiscal years, "baseline" inventories (i.e., those in the Army program database), and the new "target" inventories implied by the deltas that have been entered. The rest of the columns are for deltas.

Deltas generated by forces actions are on the right. The FSC System keeps track of deltas for different categories of force and support organizations—called "claimants"—that tend to have different equipment operating costs. Cutting the 998th Armor division created deltas only for the claimant representing active forces overseas, so there are deltas only in the first claimant column of the illustrated worksheet.

Note, however, a column near the center of the worksheet in which you can directly "Enter Inventory Deltas." (Like other editable FSC data, the entries in these cells appear in blue font on-screen, here including this column and the rows marked "ORF.")

---

[1] No warrant officer domains are used for the Air Force.

[2] Active Guard/Reserve (AGR) personnel are counted both as drill personnel and civilians because they generate pay and support costs in both roles.

[3] This worksheet only displays panels for the attack helicopters that were affected by cutting the 998th Armor division. If there are other attack helicopters in the Army inventory, you could add panels for them by pressing the "Select another item" button on this worksheet; then you could enter deltas for them.

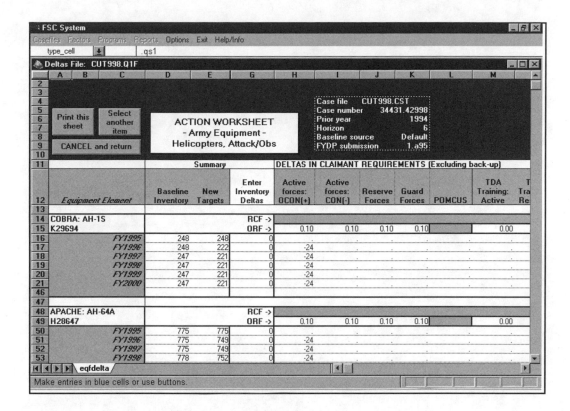

Figure 6.3 — Equipment Action Worksheet (Army Attack Helicopters)

For example, if you do not want to retire the helicopters released by units in the 998th Armor division, you could enter offsetting deltas so that the target inventories again equal the baseline inventories. If you enter deltas in this worksheet, a button allowing you to return to the casesheet and apply the revised action to your casesheet will become visible in the red header area (shown here in black).

Similar options to enter deltas directly are also available in other equipment and personnel action worksheets. In this way, the FSC System gives you control over the assumptions you make about how a forces action might be implemented. In the next chapter, we will explore more opportunities to control the methods and data used in your analysis.

Chapter Seven

SELECTING METHODS AND OTHER OPTIONS FOR ASSET ACTIONS

In this chapter, you will select among alternative options for analyzing the cost effects of deltas in assets. The illustration will pertain to the Attack Helicopter deltas associated with deactivating the 998th Armor division.

OPTIONS FOR ANALYZING ASSET COST EFFECTS

The FSC System offers several choices for analyzing asset costs. You can use either of two costing models: a simplified version called "short form," or a dynamic model called "full." You can also decide whether each action should use the default FSC factor files or custom factor files you might have listed as replacements. You can choose among options for entering constraints (such as end-strength limits) if you use the full costing method, and among options for allocating personnel or equipment shortages among claimants if constraints prevent all demands from being filled.

A newly entered action is automatically set to use default methods and assumptions. For example, all the asset actions that were constructed when you computed the Army forces action in Chapter Six were set to use the short-form method. But you can easily modify the default entries, as illustrated in Figure 7.1: simply select the action cell showing the option to be changed and press the casesheet "Edit action options" button.

For example, if you wanted to apply custom factors to an asset action, you would select the action cell under the "Data" heading, and then click the "Edit action options" button. A dialog box will ask whether you want to change the option from default factors to custom factors. (To do this, you must first complete the steps described in Chapter Two: Create and save an appropriate custom factors file, then revise the factors file listing on the casesheet to refer to the custom factors file.) If you do so, once you have closed the dialog box, the data cell in the selected action will indicate that you intend to use custom factors. You can, of course, switch back to the default factors by repeating the options editing process.

APPLYING THE FULL COSTING METHOD

When you viewed the deltas for Army Attack Helicopters, you may have noticed (in Figure 6.3) that cutting the 998th Army division would reduce the FY97–99 total in-

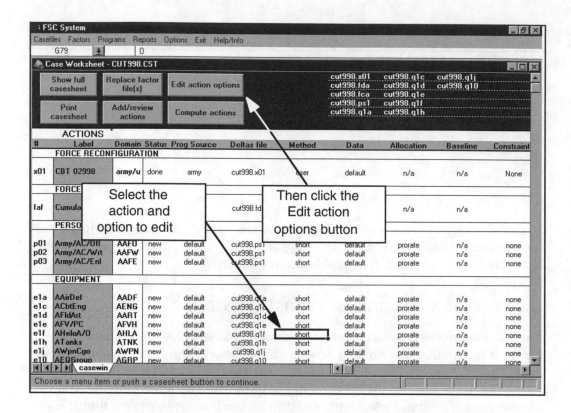

Figure 7.1 — Using the "Edit Action Options" Button on the Casesheet

ventory target for the AH-1S Cobra helicopter from the baseline of 247 to 221.[1] That reduction is large enough in proportionate terms to have a significant effect on total Army costs; therefore, you might want to use a relatively detailed, dynamic model to estimate those effects. To do so, use the full costing method.

Unlike the short-form costing method that simply applies marginal cost factors to the asset deltas, the full costing method proceeds in three stages. First, it estimates the total yearly costs of achieving and operating the inventories reported in the baseline program for the asset in question. Next, it separately estimates the total yearly costs to achieve and operate the target inventories generated by applying the deltas to the baseline program (called an "excursion" from the baseline). Finally, the cost effects are the differences between the excursion and baseline cost estimates in each year.

The baseline and excursion cost estimates are computed separately[2] and stored in separate output worksheets. Therefore, the baseline and excursion analyses are

---

[1]The active forces overseas claimant demands 24 fewer helicopters, but that also implies a reduction of two in Operational Readiness Float. The ORF deltas are computed automatically using the float factors displayed on the asset-deltas worksheet.

[2]The baseline and excursion actions can use different factor data files or be subject to different constraints. Those features are useful, for example, if the change in desired inventories is accompanied by a change in expected pay rates for personnel or fuel prices for equipment.

recorded and computed as separate actions. You can create a new baseline action on your current case, or you can borrow one you computed for a previous case. If you select an action cell under the Method heading and click the "Edit action options" button, the dialog will ask whether you want to use the short or full method; if you choose the latter, the dialog offers the options to create or borrow results from a baseline action. This dialog is illustrated in Figure 7.2.

Assuming that you decide to create a new baseline for Attack Helicopters, the new baseline action will be inserted just above the deltas action for helicopters. A baseline action is easy to distinguish from deltas actions because the "key" code in the first column of the action row begins with a number instead of a letter. This is true for personnel as well as equipment baseline actions.

CREATING ADDITIONAL BASELINE ACTIONS

If desired, you could enter additional baseline actions on this case. You would accomplish this by converting more of the existing actions to the full method. You also could add more baselines for the Army Attack Helicopters.

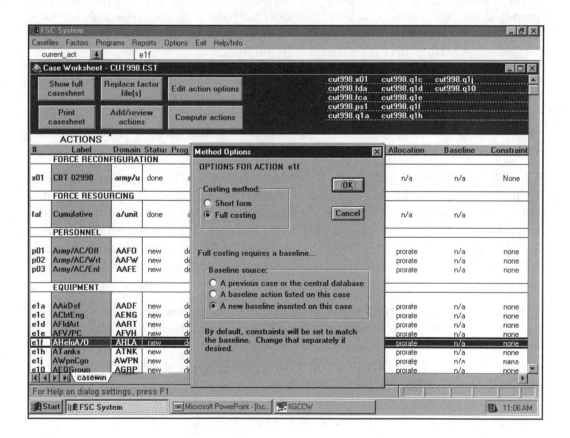

Figure 7.2 — Dialog Box for Choosing Alternative Methods

A given case can have only one asset action that has deltas within a given domain (i.e., for a given personnel category or set of equipment items)—but that is not true for asset baselines. Only actions that contain deltas result in cost effects reported on a given casesheet. Because baseline actions do not contain deltas, they do not affect casesheet cost results, and a given case can have many baseline actions pertaining to the same items of equipment or categories of personnel.

Analysts who frequently use the full method may find it convenient to devote case-sheets to computing baselines that can later be used for multiple cost analyses. To add more baselines, use the "Add/review actions" button, select one of the asset categories (and Service) in the dialog box, then select the "Calculate current program" option. Because baseline (and excursion) actions use a detailed inventory projection model, they take time to calculate. Placing several baselines on a given case and then computing the entire case in batch mode (perhaps during lunch hour) can be a convenient way to generate a lot of baseline results for future use.

EDITING OTHER ASSET ACTION OPTIONS

If you use the full costing method, you can place constraints on how the assets are managed, and you can choose how the model will handle the excesses or shortfalls in assets that the constraints generate. We will not illustrate these for the Army case, but we will describe how the constraints option works.

Suppose, for example, that you were using the full method to estimate Army commissioned officer costs, and you decided to constrain excursion personnel end-strengths to match the baseline program. In your hypothetical case, where you have cut force structure, holding end-strength constant would mean that fill rates in remaining units would rise.

The FSC System would distribute the personnel released by the deactivated units according to an "allocation rule" that you specify. The default allocation is the "prorate" rule, which redistributes the reassigned personnel to remaining units in proportion to their end-strengths; this would be done for each year of the analysis. Notably, personnel costs might rise or fall relative to the baseline program, depending on whether and how personnel are reassigned between overseas locations, where pay and other costs are higher, and lower-cost locations in the United States.

To enter constraints or modify the allocation rule, the procedure is the same as for changing cost methods. Select the action cell referring to the option you wish to change, then press the "Edit action options" button and enter your selections in the dialog box that appears. When you have made all the option selections you desire, compute your case and look at the results.

COMPLETING ALL COMPUTING

Now that you have finished entering and editing actions, you are ready to compute the entire case. Click the "Compute actions" button on the casesheet and select the batch-mode option. Because this is a large case with many actions (including some full-costing actions), it could take several minutes to compute. You can observe progress by watching the messages displayed at the bottom left corner of your monitor. After each action is computed in batch mode, the casesheet and its supporting worksheets are resaved automatically.[3]

> **TUTORIAL**: To convert the Attack Helicopters action method to full, select the cell containing "short" on the "e1f" action record. Select the "Full costing" and "A new baseline inserted on this case" options in the dialog box, and click OK. To compute all actions on the casesheet, click on the casesheet "Compute actions" button. In the dialog box, select the "Compute all actions in batch mode" option and click OK.

---

[3]At various stages during batch processing, the system will allow you to interrupt processing by pressing the escape key. You may need to press it more than once, because the system ignores the key during critical processing steps.

Chapter Eight

VIEWING RESULTS AND EXITING

Now that your case is completely computed, you can view the results and exit the FSC System.

HOW COST RESULTS ARE STORED AND REPORTED

As each action is computed, its detailed results are stored in a separate output file. The name of the file is listed in the Result column of the action record on the case-sheet, visible by scrolling to the right on the active screen. The detailed action results can be opened at any time (even when the case is closed) by making selections from the **Reports** menu.

The casesheet also records a summary of the cost effects for actions on the case. The cost deltas are shown by year for each action that generates cost effects, by major appropriation. Cost deltas are not generated by forces actions or asset baselines, but all other types of actions will generate results for the casesheet summary section.

VIEWING COST RESULTS FROM THE CASESHEET

To see the summary cost deltas on the casesheet, click the "View results" button. If you want to see both the summary and the detail costs for a particular action, follow the instructions in Figure 8.1. Select any cell on that action record before clicking "View results." Because baseline actions by themselves do not generate cost deltas, they will have no entries in the casesheet summary section. Therefore, if you select a cell on a baseline action before pressing the "View results" button, the system will ask if you want to see the detailed baseline output report. Otherwise, the FSC System will go to the summary cost section of the casesheet, and then ask if you want to see the detailed report for the action you selected.

The first delta-cost section you see after pressing the "View results" button contains cost deltas measured in constant dollars. If you scroll down, another panel showing the same results in then-year dollars comes into view.

As illustrated in the background of Figure 8.2, delta-cost summaries have separate rows for each action's cost deltas by appropriation. The entries also distinguish between the direct and indirect cost deltas for assets; direct costs are modeled under

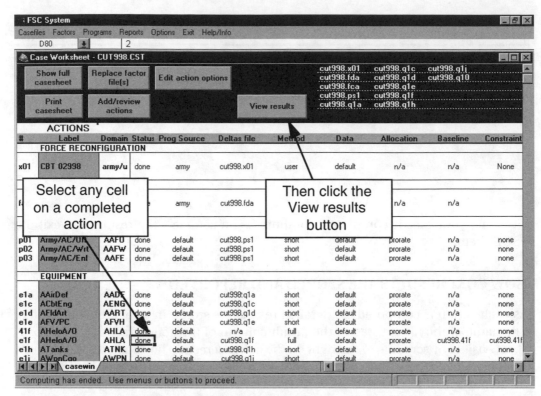

Figure 8.1 — Using the Casesheet "View Results" Button

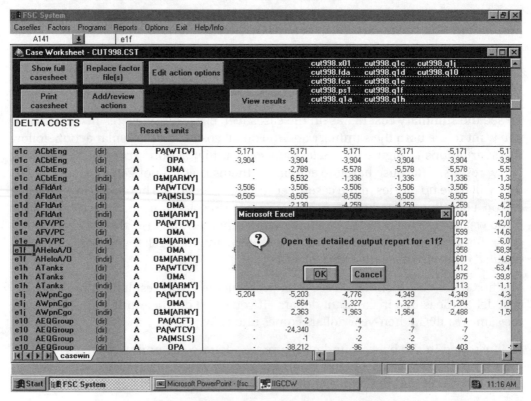

Figure 8.2 — Casesheet Delta Cost Summary (with dialog box)

either the short or full-costing method, whereas the indirect cost estimates are based on preliminary, aggregative factors.[1] Appropriation subtotals and a grand total of delta costs by fiscal year appear at the bottom of each delta-cost section.

> **TUTORIAL:** Select a cell on the deltas action for helicopters (key code e1f) and press the "View results" button. If you want to scroll through the delta-cost summary, click "Cancel" in the dialog box that asks if you want to see the detailed results report for action e1f. Then, when you have finished viewing the summary, select a cell in one of the delta-cost rows labeled e1f, press the "View results" button again, and click OK in the dialog box.

DETAILED COST OUTPUT REPORTS

A personnel or equipment cost report contains supporting detail for the estimates shown on the casesheet, as well as additional information that can be helpful for interpreting the summary cost deltas. For example, the reports show "fill rates"—the fractions of total asset targets that are met by projected inventories. Personnel output reports also indicate whether the personnel force is becoming more senior (available only with full costing), and equipment reports indicate whether depot maintenance workloads (for a given equipment item or category) are increasing over time. The equipment cost detail report for Army Attack Helicopters appears in Figure 8.3.

Using buttons on the report worksheets allows you to print the sheets or save a copy of them (in your *fsc_user* directory) or return to the casesheet. Because the casesheet and detailed output reports are simple Excel spreadsheets, copies can be reformatted or used to construct graphs for written reports concerning your analysis.

> **TUTORIAL:** Scroll through the detail report to see its contents. Then click the "Close and return" button. If desired, you may view other detail reports from your case using the same procedures.

FINISHING UP AND EXITING

Your case analysis is complete. To review, the chart in Figure 8.4 summarizes the features of your analysis. You can now consider variations on the case—different force cuts or adjustments in personnel or equipment management policies or alternative estimation options—to see how they might affect the results. For example, if you would like to consider also cutting NDC and CSS forces in echelons above division, you could follow the instructions in the Appendix.

---

[1]Future versions of the FSC System will include advanced modeling to estimate indirect costs (i.e., costs for infrastructure support). Such models would adjust the preliminary indirect cost estimates to reflect infrastructure management and pricing policies.

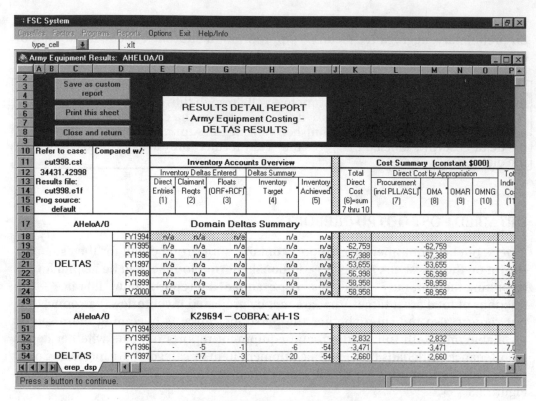

Figure 8.3 — Detailed Equipment Cost Output Report

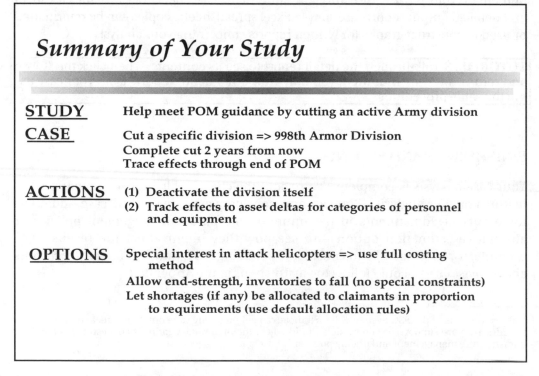

Figure 8.4 — Overview of the Completed Case Analysis

It is easy to transfer data from your case to a personal Excel worksheet. While your case file is still open, use the **Exit** menu command "To Excel" to restore the standard Excel menu bar. Open a new standard worksheet and copy data from FSC displays. When you have finished, a special command "Return to FSC" under the Excel **Files** menu allows you to return to the FSC menu bar.

When you have finished using the FSC System, choose "To Windows" ("To Finder" on a Macintosh) and both the FSC System and Excel will close.

TUTORIAL: When you have completely finished this tutorial, use the FSC **Exit** menu and select "To Windows" ("To Finder" on a Macintosh) to quit Excel entirely.

Chapter Nine
CONCLUDING COMMENTS

This manual illustrates some of the analysis choices available from the FSC System and showed you the procedures for taking advantage of them. There is much more about the system that the manual could not cover, including the system's design philosophy, methodologies, and data sources.

Inevitably, any costing methodology must make simplifying assumptions, both to overcome source-data limitations and to handle uncertainties about how decision-makers might respond to budget contingencies. The FSC System's models reflect empirical analysis concerning these issues and are designed to handle them in consistent and realistic ways. Beyond that, the system is designed to permit you, the analyst, to apply your own knowledge and expertise to the issues you investigate. Ultimately, you must manage the data and methods available through the FSC System to generate meaningful findings. The system is intended to help you, not replace you.

We encourage you to learn more about the system in two ways. One is learning by doing; as you perform costing exercises and examine their results, observe patterns in the way the system models forces, assets, and infrastructure that will enrich your understanding of the data and costing formulas. Also, refer to other FSC System documentation, including online Help and other user and technical guides in this series.

USING A FORCES TEMPLATE

A forces **Template** is a listing of the units or assets comprising a *notional* component of the force structure, such as a typical heavy division or the NDC and CSS organizations that might support a division when it is deployed.

Because templates refer to notional structures rather than to organizations found in the current forces program, templates are considered to be "factors." The FSC System's factors database includes several templates that you may use to modify the force structure in your cases. The FSC templates are listed in the Factors Files section of the casesheet and can be viewed by using the **Factors** menu.[1]

In a case analysis, you can enter deltas in forces by applying templates to the currently programmed force structure. That is, you can add (activate) or subtract (deactivate) the template units or assets. The procedure for applying templates to create forces actions is illustrated in this appendix.

Templates have three features that affect the way they may be used in an FSC case analysis:

• **Two distinct types of templates are available.** "Unit templates" list *unit counts* for direct combat forces; "asset templates" list *asset quantities* for categories of combat-supporting forces. You can apply *unit* templates only to the portions of the force structure that are measured in units (such as battalions in a division or aircraft squadrons in a wing), and you can apply *asset* templates only to the portions of the force structure that are measured in asset quantities (such as Army NDC and CSS organizations or Air Force base operating support organizations). Therefore, if you intend to apply templates both to direct and supporting combat forces, you will need to enter separate actions (using separate templates) for them.

• **Templates may refer to units or resources that are not in the current forces program.** If you apply a template as an addition to the current force structure, the FSC System will add the template's units or asset quantities even if there are no other "like" structures in the current program. However, if you apply a template as a way of cutting forces, the FSC System *will only be able to deactivate the tem-*

---

[1]For example, to view any of the Army templates, select "Army force elements" from the Factors menu and then choose "Force Templates" in the dialog box that follows.

plate's elements to the extent that "like" elements are found in the current program for each fiscal year. For example, if a template referred to tank battalions equipped with M60 tanks and no such units existed in the program (e.g., if all programmed tank battalions had M1A1s or M1A2s), then the FSC System could *not* deactivate the template's tank battalions. Consequently, it is important to use templates only as an *aid* to creating force actions. Force revisions entered by a template should be carefully checked and revised as needed, especially when templates are used to *cut* forces.

- **Templates list units or assets without regard to component or region.** When you apply a template, the FSC System asks you to specify *preferences* for applying the deltas across components and regions. The default preference is to prorate the deltas, i.e., to allocate them in proportion to the configuration of the currently programmed force structure. (For example, if all the programmed forces in the Engineer branch of CSS were located in CONUS but half were in the Reserve and half were in the Guard, the default option would apply half the template asset amounts to the Reserve Engineer branch in CONUS and half to the Guard Engineer branch in CONUS.) If you set other preferences, however, the FSC System attempts to apply the deltas to the specified component(s) and region(s)— but if the deltas are cuts that are too large to be absorbed by the currently programmed force in the specified component and region, the FSC System will not be able to apply all the cuts to the specified forces.

To apply a template to the programmed force structure—i.e., to use a template to help create a forces action in a case—follow these steps:

Step 1. Open an existing case or start a new one. Then initiate a forces action in the usual way, choosing *Forces* and the desired Service in the "Add/review actions" dialog box.

Step 2. Specify a forces extract as you would for any forces action. When you intend to apply a template, however, you might find it helpful to use the *Template-related orgs* option. This tells the FSC System to extract the portion of programmed forces that corresponds to the organizations represented in the type of template you specify.

As an illustration, suppose that you intend to cut some NDC and CSS assets that might not be needed if the 998th Armor Division were no longer in the force structure; specifically, suppose you intend to cut assets in Echelons Above Division (EAD) that might support the 998th if it were deployed to Europe or Southwest Asia (SWA). Figure A.1 shows the appropriate extract selections.

The extract based on a template group lists the *total* asset quantities in the current Army program for each of the NDC and CSS categories listed in the specified template. Thus, the extract shows you the maximum assets that could be cut from the relevant portion of the force structure. An illustration of the EAD extract appears in Figure A.2.

Figure A.1 — Dialog for Selecting a Template-Related Forces Extract

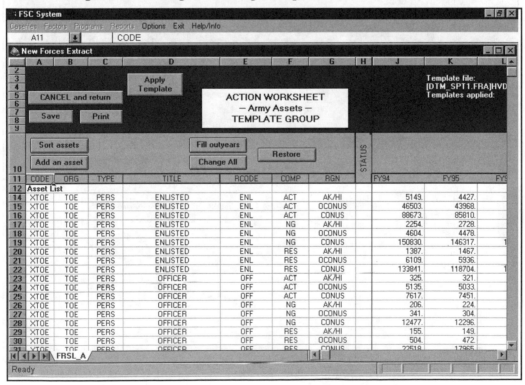

Figure A.2 — Extract of Forces in Echelons Above Division

Step 3. Press the "Apply Template" button on the action worksheet. The button opens a dialog box that asks for more information about which template you wish to apply. You can select one of the templates stored in the FSC central database or one that you customized by following the same procedures used to customize any factor file. See Figure A.3.

Step 4. Indicate how you wish to apply the template. The dialog box in Figure A.4 asks whether and when the template's forces should be added to the programmed force structure or removed from it. You are also asked for your preferences regarding the component(s) and region(s) to which the template deltas should apply. This example assumes you intend to cut NDC/CSS forces in FY96 by using the default preferences that prorate the cuts across components and regions.

To apply the template, the FSC System analyzes each element (each row of data) in the template to determine how it should be allocated across components and regions in each year. This takes time. Messages at the bottom of the screen indicate progress. When the procedure is completed, the extract display (illustrated in Figure A.5) shows in boldface the force revisions that have been made, just as though you had entered them yourself.

Step 5. Examine—and, if necessary, correct—the force edits that result from applying the template. It is especially important to evaluate any *negative* entries that may

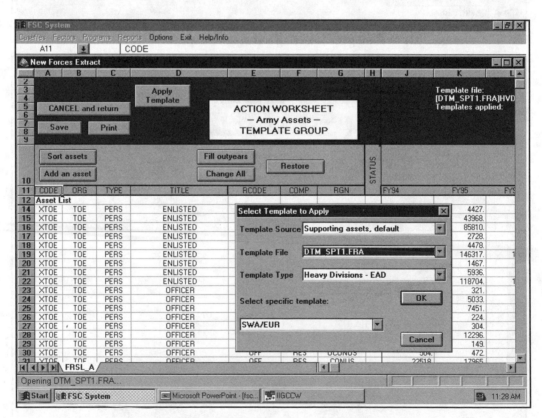

Figure A.3 — Selecting a Template to Apply

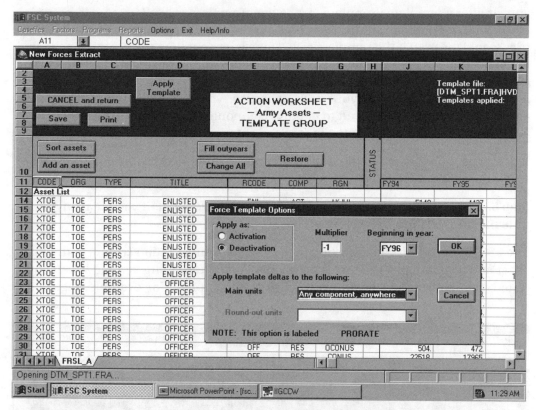

Figure A.4 — Applying the Template

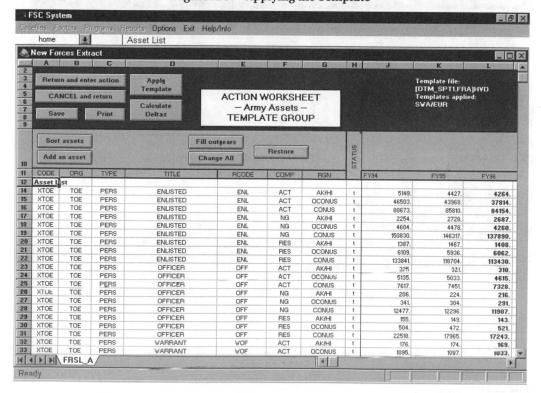

Figure A.5 — Forces Worksheet After Application of Template

appear after a template is applied. Since the "Template-related orgs" extract begins by showing *all* the relevant units or assets in the Army force structure, negative entries would mean that the template called for a larger cut than the programmed force could absorb. Therefore, the FSC System will not apply those entries in full. (The system will automatically reset the negative entries to zero before computing the forces action.) Based on your own expertise, you might decide that negative amounts should be transferred to another component or region.[2] If so, make those adjustments before returning to the case and entering the action.

In this illustration, there are no negative entries. Consequently, it is not necessary to revise the entries.

Step 6. Return to the casesheet to enter the forces action and (if you are ready) compute the case.

The FSC System enters a new action in the Force Reconfiguration section of the casesheet and accumulates the forces deltas in a Force Resourcing action. When computed, the latter action creates or revises asset actions. For example, Figure A.6

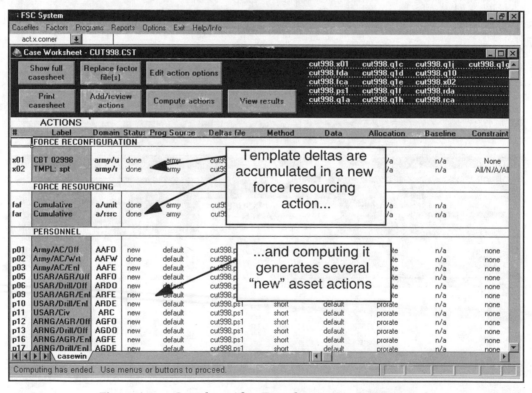

Figure A.6 — Casesheet After Template Action Is Computed

---

[2]When the template entries are prorated, it is impossible to eliminate the negative entries by reallocating them to other components or regions —the cuts are simply too large in the aggregate. However, reallocating the negative amounts could be a solution in cases that apply the templates according to other preferences.

shows the addition of several "new" personnel actions after the Force Resourcing action is computed.

Some of the asset actions marked "new" may, indeed, be newly created. In the illustrated case, the initial forces action (cutting the 998th Armor Division) only affected active component personnel whereas the NDC/CSS cuts cause deltas for Reserve personnel as well. Other actions are labeled "new" because the template action modified the asset actions that had previously been costed. The FSC System recognizes that those action(s) must be recalculated using the new deltas; they are marked "new" to prepare them to be recomputed.

The system keeps track of the previously computed personnel and equipment actions. It lists them in the Case Modification Log as illustrated in Figure A.7. Notice that the log entries include the names of results files that were previously created for each action. Those results can still be viewed from the **Reports** menu and are stored as part of this case. However, those results are no longer included in the cost deltas summarized on this casesheet.

When the "new" actions are computed, their results are added to the cost deltas shown on the casesheet. Thus, the modified casesheet summarizes the combined

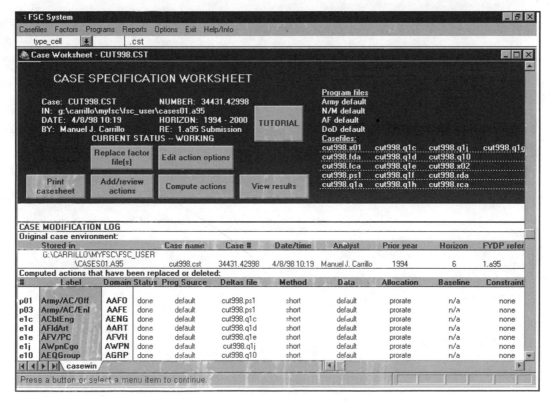

Figure A.7 — Entries in the Case Modification Log

cost effects of the initial forces cut (e.g., the 998th Armor Division) and those entered with the assistance of the forces template. You can view the cost summary and the detailed results underlying it by using the same techniques that were illustrated in the main body of this manual.